FOR THE LOVE OF VIRGINIA BEEF

By Angela White Ball

Illustrated by Grayson B. McMurray

For the Love of Virgina Beef
Written by Angela White Ball
Illustrated by Grayson B. McMurray
Background images: Adobe Stock
Published 2026
Skippy Creek
Imprint of Jan-Carol Publishing, Inc.

ISBN: 978-1-970471-42-7 (Paperback)
ISBN: 978-1-970471-43-4 (Hardcover)
Library of Congress Control Number: On file

You may contact the publisher:
Jan-Carol Publishing, Inc.
PO Box 701
Johnson City, TN 37605
publisher@jancarolpublishing.com
www.jancarolpublishing.com

Jan-Carol Publishing, Inc
"every story needs a book"

Dedicated to:

My father, Allan Paige White (1937–2025), the first farmer who blessed my life and who taught me the value of hard work and the beauty of Virginia pastures.

My favorite farmer and greatest love, my son, Collin Reese Ball, who carries the farming legacy forward. This book is for you and every Virginia farmer who turns dreams into fields of promise.

The K-12 teachers, who sow seeds of curiosity and watch them grow into lifelong learning. Among them, my mother, Susie White, whose work in the classroom helped grow generations of young readers and writers.

A Note from the Author

Hi, friends! I wrote ***For the Love of Virginia Beef*** to help you learn all about how beef cattle are raised in my home state of Virginia. This book takes readers on a journey through every stage of raising beef cattle — so it's a little longer because there is so much to share. Along the way, you'll discover the top 25 beef-producing counties in Virginia.

Key Features:

- **Read-Aloud Friendly:** Ideal for Pre-K through 5th-grade classrooms; approximately 30 minutes.
- **Educational Value:** Offers a true-to-life depiction of Virginia's beef cattle industry.
- **Interactive & Flexible:** Built-in activities and natural stopping points make it easy to adapt for different age groups and time frames.
- **Stretch & Moo Breaks:** Fun pauses throughout the book encourage movement and allow readers to resume later if needed.
- **Moo-vement Moments:** Short, playful activities to keep young readers engaged and active.
- **Beef Up Your Knowledge:** Extra facts for curious readers who want to dive deeper.

This book is designed to be **informative, interactive, and adaptable**, making it a valuable resource for educators and families.

You can read the whole book in about 30 minutes or take your time and stop at the breaks. Either way, I hope you have fun learning about Virginia's beef cattle industry!

We're traveling across the beautiful
Commonwealth of Virginia
to explore beef cattle farms.

Grab your favorite boots,
tip your cowboy hat,
and hold on tight.
We're off to say **"Howdy!"**
to Virginia's fabulous farmers
and meet their **moo-velous** friends.
Yee-Haw!

From the Appalachian Mountains
to the sandy Eastern Shore,
for the love of beef,
Virginia farmers work hard
through winter, spring,
summer, and fall
caring for their cattle
from sunrise to sunset.

As part of our farm adventure,
we're hopping into a time machine —
ZOOM! Back we go to Jamestown, Virginia,
the very first English settlement.
Life was tough for those early colonists
as they learned to live in a brand-new world.

But guess what?
By 1610, cows were here too — moo-velous news!

For the love of beef, cattle became super important!
They gave meat for hearty meals and milk for creamy treats.
Fresh milk was churned into golden butter
and aged into cheese. **Yum!**
Cattle hides were crafted into leather
for clothes and everyday tools.
And the biggest cows?
They helped plow fields and pull heavy carts.
Talk about hard-working helpers!

So, what are beef cattle? Moo-hoo-hoo!
They're special bovines (cattle)
with a job to do.

Farmers and ranchers
raise them with care for tasty meat —
what a treat!
And now…drumroll, please…
That meat has a name. It's called BEEF!

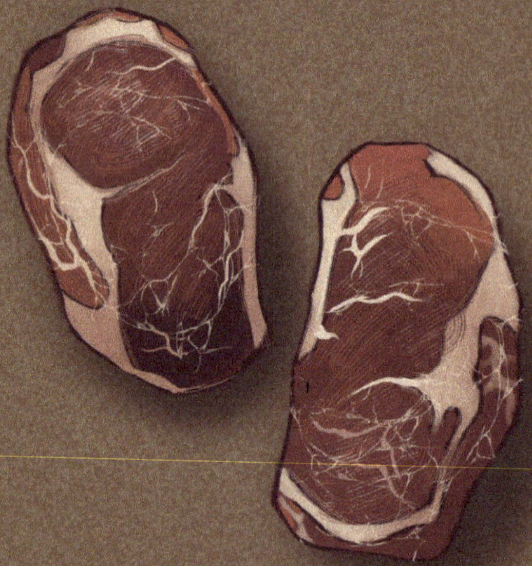

Moo-vement Moment

Can you say "BEEF" really loud?
Go ahead — shout it out!
BEEEEEF!

BEEF UP YOUR KNOWLEDGE

Bovine: Noun; comes from the Latin word "cow"; any animal of the subfamily Bovinae including cattle, bison, and oxen

Beef: Noun; flesh of mature or almost mature cattle used as food

For the love of beef, you can find beef cattle roaming in every county in Virginia.
They're not all the same — oh no, not at all.
Some are short, some are tall. Colors galore — black, red, and white — what a colorful sight!
Some are long and lean, others round and wide, each with a breed that shows their pride.

On average, a beef cow (female) weighs 1,200–1,400 pounds, while an average beef bull (male) weighs 1,500–2,200 pounds. **WOW**, that's heavy!

There are more than 70 breeds across the land, but in Virginia, three take a stand: Angus, Hereford, and Simmental too. Plus Charolais, Limousin, and Shorthorn are in view.

Moo-vement Moment

Can you moo like a cow?
Go on — give it your loudest MOOOO!

BEEF UP YOUR KNOWLEDGE

Breed: Noun; group of animals with shared, identifiable characteristics. A breed is like a branch of a family.

Example: All dogs are one species or big family. Within the big dog family, there are numerous breeds or smaller families like Australian Cattle Dogs, Boston Terriers, Cocker Spaniels, and Golden Retrievers.

For the love of beef, Virginia farmers help keep the economy moo-ving!

Millions of dollars flow into the state each year thanks to beef cattle.

In fact, raising beef cattle is Virginia's second-largest agricultural business. **WOW!**

And here's something special:
About 90 percent of beef cattle farms are family-owned.
Moo here, moo there — farm families everywhere!

Families work together on two types of beef farms:

- **Cow-Calf Operations**, where baby calves are born and raised.

AND

- **Stocker Operations**, where young cattle grow big and strong.

BEEF UP YOUR KNOWLEDGE

In Virginia, there are cow-calf operations and stocker operations. Some farms have both types of operations.

Cow-Calf Operation:
The farm has a herd of beef cows and bulls that breed to produce calves annually. The calves are raised until weaning. Weaned calves are sold to other producers or feedlots.

Backgrounding or Stocker Operation:
Weaned calves are grown on pasture or supplemented feed until they reach 700-900 lbs. These stockers are sent to a feedlot to reach market weight before going to harvesting or processing facilities.

As our adventure continues, the sun peeks over the horizon, painting the sky pink and orange.
Across Virginia, cattlemen and cattlewomen call out:
"Rise and shine — it's time to get to work!"

January opens the gate to a brand-new year, with winter's icy magic all around the farm.

For the love of beef, a farmer in Russell County turns the key — **VROOM!** His big red truck roars to life. The air is crisp and cool inside the cab.

He presses the gas — **ZOOM!**
But — **WHOA!** — the tires slip on a thin dusting of snow.
Where's he headed? To the barn, of course!

It's time to check the calving supplies, because January is perfect for getting ready for baby calves.

Feeding Tube — **check!**

Ear Tags — **check!**

Animal Health Supplies like Ob Chains, Syringes, and Vitamins — **check!**

Gloves — **check!**

Calving Book — **check!**

And don't forget the Bedding — soft straw and fluffy shavings — **check, check!**

Everything a farmer needs to welcome new life on the farm.

Moo-vement Moment

Can you count how many checks we just said? (Here's a hint: it's more than five!)

BEEF UP YOUR KNOWLEDGE

In Virginia, beef cows calve during two main seasons: spring and fall.

Spring Calving Season: Approximately January to March

Fall Calving Season: September to November

For the love of beef,
a cattlewoman in Augusta County bundles up —
coveralls, muck boots, and a purple beanie —
ready to brave the chilly night.

She hops into her gray side-by-side — **VROOM!** —
and zooms toward the barn lot.

Why? Because it's finally calving season in Virginia! January, February, and March bring bouncing baby calves to farms across the state.

Farmers work both day and night, checking on cows who are waiting for their little ones.

What's a cow? She's an adult female, a girl, who's already had at least one calf in her lifetime.

A cow carries the calf in her belly for about 283 days. That's nine months, just like human moms.

And when the big day comes, a calf is born — a baby, boy or girl, usually weighing between 75 and 100 pounds. That's heavier than most kids in kindergarten! Moo-velous, right?

Moo-vement Moment

Can you pretend to rock a baby calf in your arms?

BEEF UP YOUR KNOWLEDGE

Cow: female bovine that has grown up and had at least one baby (calf)

Bull: male bovine that is not castrated and has grown up to be used for breeding

Calf: a baby bovine (male or female) that still drinks milk from its mother

Heifer: young female bovine that has not had a baby

Steer: male bovine that has been castrated; used for meat, not breeding

Castrated: an operation that guarantees a male calf cannot make a baby with a cow or heifer

Gestation: period of time when a baby calf grows inside the mother's belly from a tiny speck into a fully formed calf

For the love of beef,
on a frosty February morning,
a **Washington County** farmer cranks up
his mighty green tractor — **VROOM!** —
that is hooked to a big orange piece of equipment
called a mixer. It's feeding time!

Winter's chill means the grass won't grow,
so the mamma cows need a tasty meal
to stay strong and make milk for their calves.

What's on the menu?
A high-energy, high-protein feast!
The tractor pulls the mixer as it spits out a mix of hay and corn silage —
WOOSH!

And look out! Here they come!
The mamma cows trot, then run,
tails swishing and noses sniffing.
Breakfast is served!

BEEF UP YOUR KNOWLEDGE

Hay: Type of livestock feed made of green grass and plants that farmers cut, dry in the sun, and bundle into large bales, which are heavy rolls or blocks. Hay is a food source packed with energy and vitamins when fresh grass is not available.

Corn Silage: Type of livestock feed made from the whole corn plant – stalk, leaves, ear of corn. Ears of corn are chopped up and turned into a pickled salad of sorts.

March means spring is here — well…almost!
In Virginia, the wind still whistles,
and the air can feel chilly.

For the love of beef, a farmer in Bedford County
hops into her sturdy white feed truck —
VROOM! — and rumbles down the lane.

What's a feed truck? It's a super-special truck
that feeds cows in the pasture.

She drives the truck
beside a row of concrete troughs —
CLUNK! CLUNK! —
and grain shoots out of the bin —
SWHOOSH! — filling each trough
with yummy pellets of grain.
This clever truck even measures
the perfect amount for every cow.
Dinner is served!

For the love of beef, a Pittsylvania County farmer loads up his camo-colored side-by-side with tools and fencing gear — **CLANK! CLANK!** It's time for a big job.

The farm crew zooms off to check miles of fence. Are there fallen trees? Any big holes?

Everything must be safe and sound. Why all the fuss? Because it's almost time to turn the cattle out on the fresh green pastures. Moo-velous days ahead!

Moo-vement Moment

Let's pretend to hammer a post.

STRETCH and MOO! Break

It's time to stretch and moo like a cow!
Stand up tall like a big, strong cow —
stomp your feet three times!
Now, moo as loud as you can: **MOOOO!**

Ready to hit the trail again? Saddle up!
Now take a big breath of fresh air.
Are you ready for the next part of our beef adventure?

Yee-Haw! Let's ride!

Look out across Virginia — from **Lee County** to **Fauquier County** — and you'll see calves everywhere! Black, red, and white coats dot the farm landscape like sprinkles on a cupcake.

Moo here, moo there — spring is in the air!

Moo-vement Moment

Can you hop like a calf playing in the field?
Ready? Boing, boing — MOO!

April tiptoes in with breezy days,
showering soft rain across the fields.
Pastures are greening up,
ready to burst with vegetation.

For the love of beef, cowboys and cowgirls
in **Wythe County** gather at the barn,
adjusting leather straps and patting their trusty horses.
It is time to saddle up. Today's the big day —
the cattle are heading out to pasture.

Cows are amazing grazers. Fresh green grass helps
them grow strong bones, build big muscles, and gives
them plenty of energy to moo-ve.

Working together, the crew rides out on horseback with cattle dogs leading the way. Cows with coats of black with calves by their side trot toward lush green fields.

Then — **Slurp, Slurp!** — those long black and pink tongues curl around tender blades of grass. **Yum!**

Did you know a cow can eat up to 35 to 40 pounds of grass every single day? That's A LOT of salad!

BEEF UP YOUR KNOWLEDGE

Ruminant: Cattle, sheep, and goats are ruminants. Ruminants have a stomach that allows them to eat and digest all types of vegetation and convert those into energy. The ruminant stomach consists of four compartments: rumen, reticulum, omasum, and abomasum.

Cattle can eat grass, hay, corn stalks, and other plant materials. They can eat vegetation that humans and other animals cannot. These are called forages.

Forage refers to grasses and other plants that are grazed or browsed by animals.

May dances in with sunshine, longer days,
and flowers blooming bright.
A new star is ready to take the pasture stage!

For the love of beef, a farmer in **Scott County** backs her shiny silver trailer into the pasture — **BEEP, BEEP**! The door swings open, and out steps the king of the pasture…a bull! The big black bull snorts — **SNORT!** — and bellows — **ROAR!** — as he surveys his new home before joining the cows.

What's a bull? He's an adult male — the strong father of the farm. Bulls are much larger than cows, with bigger heads and thick, muscular necks.

In Virginia, bulls join the herd in May and June as part of the breeding program — a super-important job on a cow-calf farm. Without bulls, there'd be no baby calves! Moo-velous teamwork!

Moo-vement Moment

Can you stomp your feet like a big bull?
Stomp, Stomp

For the love of beef, a farmer in **Rockingham County** climbs onto his cherry-red tractor with giant dual wheels — **VROOM!** — and pulls a wide green corn planter across the field.

What's he planting? Field corn!
Tiny seeds will tuck into the soil and sprout into emerald stalks.
By summer's end, those stalks will wear tassels of purple and yellow — just like party hats.

When harvest time comes, the corn will be gathered as crunchy ears or chopped into silage — a yummy winter treat for hungry cows.
Moo-velous farming magic from seed to feed!

Summer is finally here!
June, July, and August keep farmers busy, busy, busy.
There's so much to do — making sure the cows stay healthy and strong while caring for the land they call home.

For the love of beef, from sunrise to sunset, the work never stops, because happy cows mean moo-velous beef!

Moo-vement Moment

Can you stretch your arms like the big summer sun?
Now say it loud: "Hello, summer!

For the love of beef, a farmer in **Rockbridge County** checks every water trough — **SPLASH!** — to make sure they're full of fresh, cool water.

Why? Because cows drink A LOT. A beef cow can consume up to 30 to 50 gallons of water a day. That's as much water as a whole bathtub!

When summer turns hot and humid, cows need even more water to stay healthy and happy.

On this farm, a sparkling spring bubbles up from the ground, sending a rippling stream across the pasture. But guess what? The stream is fenced off so the cows can't splash in it.

Why? Because farmers work hard to protect this natural resource, keeping the water clean for everyone living in the watershed. Moo-velous care for the cattle and the land!

Virginia's Three Watersheds

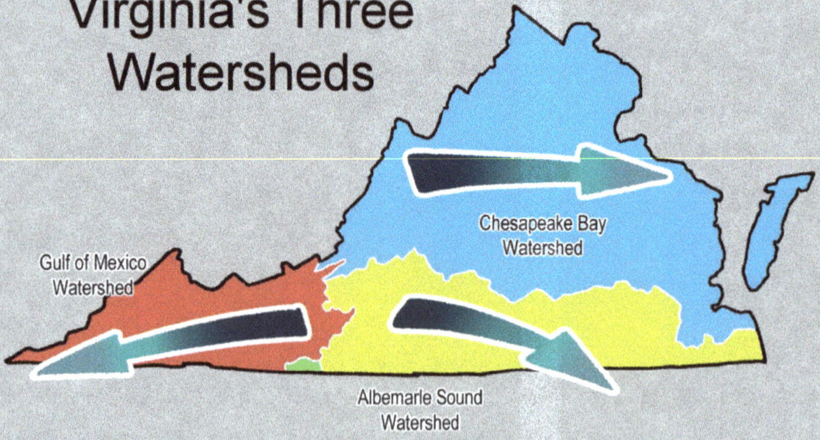

BEEF UP YOUR KNOWLEDGE

Watershed: Noun; a geographic area over which surface water (and the materials it carries) flows to a single collection place; land that water flows across or through on its way to a stream lake, wetland, or other body of water.

The watershed systems in Virginia lead to three main bodies of water. The three major watersheds in Virginia include:

Chesapeake Bay Watershed: Approximately 60% of Virginia's land area drains into the Chesapeake Bay via the James, Potomac, Rappahannock, Shenandoah, and York Rivers.

North Carolina Sounds Watershed: Also known as the Albemarle-Pamlico Sound Watershed, both of which are in North Carolina. Includes drainage from rivers in Southeastern and South-Central Viriginia (Chowan, Roanoke, Blackwater, and Nottoway Rivers).

Gulf of Mexico Watershed: Rivers in Southwest Virginia (Holston, Clinch, Powell) drain into the Tennessee River system. The New River and Big Sandy River drain into the Ohio River system. Ultimately, all these waterways drain to the Mississippi River and the Gulf of Mexico.

Summer chores aren't just about grass and water. Farmers must also keep their herds healthy. For the love of beef, a farmer in **Shenandoah County** welcomes the large animal vet — **HELLO! HELLO!** What's the plan? A herd health check-up!

Vaccinations are a big part of keeping cattle healthy, just like the shots kids get at the doctor's office. The vet gives all the beef cattle shots. These special shots protect cattle from germs like bacteria, viruses, and even tiny protozoa. Moo-velous care for happy, healthy herds!

Moo-vement Moment

Give yourself a big pat on the back — just like the farmer and vet do after a job well done!

BEEF UP YOUR KNOWLEDGE

Large Animal Veterinarian: A veterinarian, or vet, is a person qualified and authorized to practice veterinary medicine. A vet is responsible for the diagnosis, treatment, and prevention of medical conditions of animals. A large animal vet specializes in large animals to include farm animals (like cows, sheep, goats, and pigs), equine (horses), and zoo animals. A large animal vet may have an office but also visits farms.

Cattle Vaccinations: a type of medicine given to cattle to keep them healthy and protect them from getting diseases that could make them very sick, stop them from growing, or even cause death; typically, in liquid form and given via a needle (shot)

For the love of beef, a farmer in Grayson County hops on his bright yellow ATV — **BEEP! BEEP!** — and zooms across the rolling pasture.

Today's job? Rotational grazing! What's that? It means the big pasture is divided into smaller fields called paddocks, all fenced in to keep things neat and tidy.

The farmer drives through paddock after paddock until he reaches the one with the cattle.

Uh-oh...The grass is only four inches tall! Time to move the herd to a fresh green buffet.

The cows know the drill. They see the ATV and — **MOO!** — here they come, trotting right along. Through the open gate they go, happy to munch in a brand-new grassy paddock.

This smart system keeps pastures healthy and spreads out the manure. What's manure? Well, that's just a fancy word for cow poop!

Moo-velous teamwork between cows and farmers!

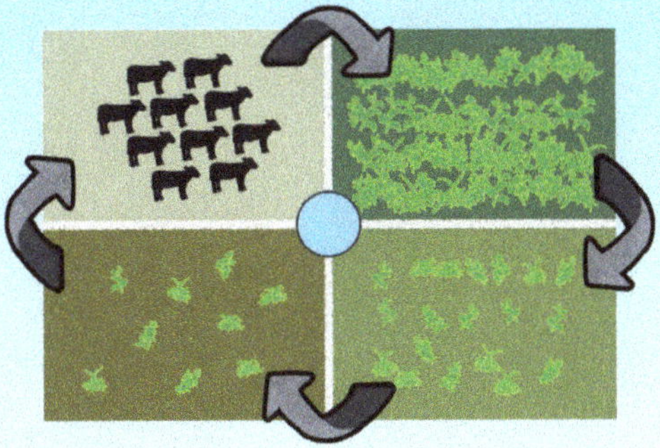

Moo-vement Moment

Can you say "manure" three times fast without giggling? Ready? Manure, manure, manure!

BEEF UP YOUR KNOWLEDGE

Rotational Grazing: A type of livestock management system where a large pasture is divided into smaller areas called paddocks. These paddocks are created by installing a fence. Livestock are moved from one grazing paddock to another. Grazing is contained in smaller areas for shorter amounts of time, which prevents overgrazing and disperses manure more evenly. When livestock are moved to a new grazing paddock, plants in the other paddocks are resting and growing. There are numerous benefits to rotational grazing:

- Plants rest and grow deeper roots, which results in greater and better forage production.
- Improved soil health due to manure being distributed more evenly. Soil organic matter (living and once living) is increased with better soil structure and moisture retention.
- Livestock health is improved with less exposure to parasites.

Summer skies aren't just for birds and butterflies — they're for smart farming too. For the love of beef, look up — way up! A drone is buzzing over a pasture in **Carroll County** like a tiny flying helper.

What's it doing? It's on a mission to find weeds. Weeds are troublemakers! They steal space and nutrients from the grass that the cows love to munch. So, the drone spots the weeds and — **ZIP!** — sprays just the patches that need it. Smart farming from the sky. Moo-velous technology at work!

Moo-vement Moment

Can you make a buzzing sound like a drone? Ready? BZZZZZZ! Now add a quick ZIP!

Summer health checks continue.
For the love of beef, a farmer in **Smyth County** returns from the vet clinic with jugs of purple pour-on medicine called dewormer. What's it for? To keep pesky parasites, like worms, away!

First, the farmer figures out the right amount for each animal based on its weight. One by one, the cattle walk down the chute — **CLUNK, CLUNK** — and stop at the head gate.

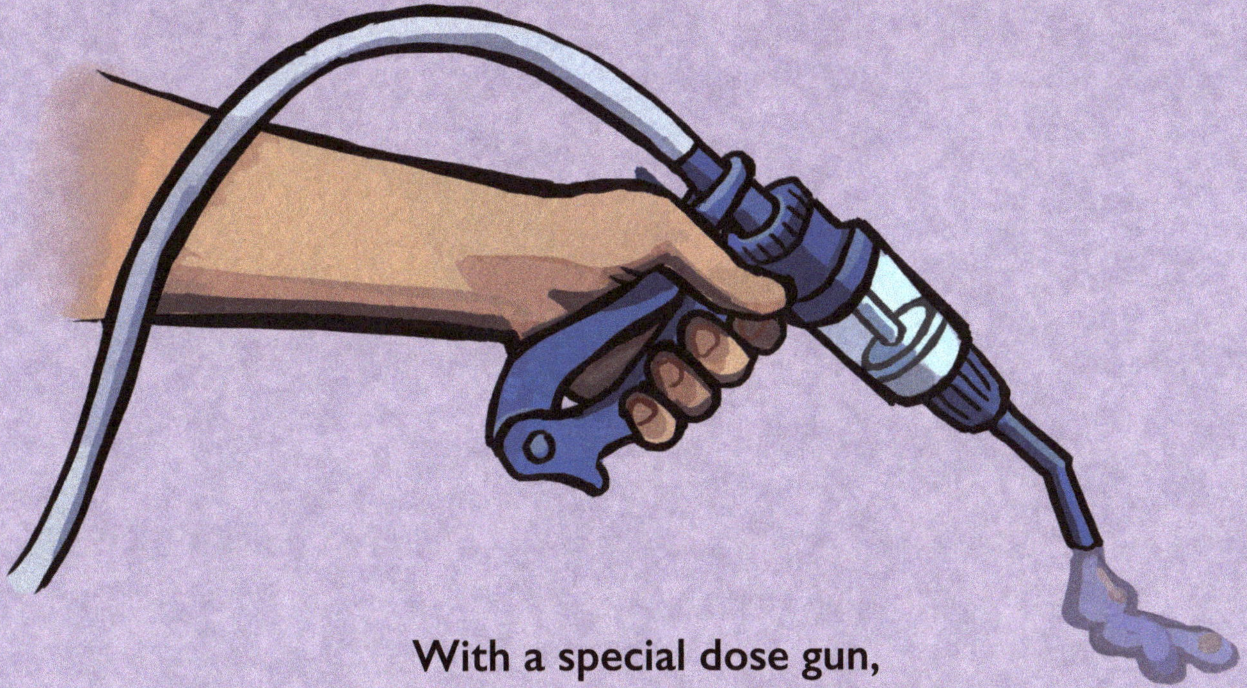

With a special dose gun, the farmer pours the purple liquid down each cow's back — **SQUISH!** — and just like that, the cows are protected from parasites and ready to moo-ve on.

Summer nutrition includes more than green grass. For the love of beef, a farmer in **Franklin County** empties a big bag of minerals into a feeder in the grazing paddock — **SHAKE, SHAKE!** Why minerals? Because cows need them to stay healthy, just like you might take a multivitamin.

Some minerals come from the plants cows eat, but others need a little extra help. So, farmers make sure their cattle get the perfect mix for strong bones, shiny coats, and happy moos.

Moo-vement Moment

Can you shake like a bag of minerals?
Ready? SHAKE, SHAKE, SHAKE! Now give a big MOO!

Summer sunshine means it is haymaking time. For the love of beef, a bold blue tractor in Pulaski County pulls a yellow hay baler across a flat field — **CLUNK, WHIRR, THUMP!** Out pops a giant round bale of hay. One bale…then another… Hundreds more dot the field like giant golden marshmallows!

When the weather is just right, farmers cut the hay, rake it into neat rows called windrows, and roll it up tight into bales.

Later, those bales will be hauled away and stored for winter, because even cows need a cozy snack when the snow starts to fall.

Summer evenings are moo-velous! The sky looks like cotton candy. For the love of beef, a cattleman in Albemarle County rides his red roan Quarter horse through the pasture. What are the cows doing? Most are relaxing — just chewing their cud — while calves race and play across the field.

What's cud? It's like a cow's chewing gum. A mix of food they've already swallowed and burped up for another chew.

Why? Grass is tough stuff, full of fibers called cellulose. Cellulose is the tough material that makes up plant cell walls. Humans can't digest it, but cows sure can, thanks to this clever trick. That burped-up cud is about the size of a baseball. And guess what? Cows can spend up to eight hours a day chewing it.

STRETCH and MOO! Break

It's time to stretch and moo like a cow!
Stand up tall like a big, strong cow —
stomp your feet three times!
Now, moo as loud as you can: **MOOOO!**

Ready to hit the trail again? Saddle up!
Now take a big breath of fresh air.
Are you ready for the next part of our beef adventure?

Yee-Haw! Let's ride!

Fall is here! September, October, November —
three colorful months across Virginia's farmscapes.
Leaves turn golden yellow, fiery red, and rusty orange.
The air feels crisp. The wind blows.
Leaves swirl and dance — like nature's own parade.

And just like trees that change their clothes,
cattle life moo-ves as the season goes.

<u>Moo-vement Moment</u>

Can you twirl like a falling leaf?
Spin, spin, spin!
Now shout it out: "Hello, fall!"

For the love of beef, on a farm in **Campbell County**, mamma cows and calves go their separate ways. It's called weaning — oh, what a change! No more milk from Mom, oh dear, but the farmer makes sure there's plenty near: lush green grass, sweet hay to munch, and calf-starter pellets for a protein punch!

For 45 to 60 days, these calves learn to stand tall, living on their own — no moms at all. There are two roads ahead for all weaned calves. Some will join a breeding program to be part of a cow-calf operation, while others will enter the stocker phase.

Moo-vement Moment

Can you stand tall like a calf learning to be strong? Stretch up high! Now say it loud: "Grow, calves, grow!"

BEEF UP YOUR KNOWLEDGE

Weaning: separation of a calf from the mother cow, which ends milk consumption and requires the calf to become independent, eating pasture, forages, or grains

- Calves are weaned around six to nine months of age.
- Weaned calves weigh 400–650 lbs.
- A recommended management practice is to wean calves 45–60 days before selling and to provide additional vaccinations at this time.

For the love of beef, a cattlewoman in **Culpeper County** strolls through her herd of weaned heifers. Notebook in hand, she scans the scene, looking for the best and brightest.

A *moo-moo* here, and *swish-swish* there, the heifers stand proud with curious stares. These young gals shine bright. They will grow and grow with all their might. One day, they will be moms with calves brand new, keeping the farm life fresh and true.

For the love of beef, a farmer in **Madison County** and another in **Appomattox County** purchase the weaned calves that remain.

Hip-Hip-Hooray!

Guess what? They're entering the backgrounding or stocker phase. Calves in this phase are called stockers. What's a stocker? Think of a cow in middle school — too big for Mom, yet not quite grown-up. They're the teenagers of the cattle herd, about six to ten months of age. *Chomp-chomp* grass! *Crunch-crunch* hay! They munch and munch to get bigger every day.

Moo-vement Moment

Now shout it out: "Grow, stockers, grow!"

BEEF UP YOUR KNOWLEDGE

Backgrounding or Stocker Phase: Weaned calves will be kept until they reach approximately 900 lbs. and can be sent to a feedlot.

Stocker Cattle/Stockers: Weaned cattle; younger (six-eight months of age) and lighter (weighing between 350–550 lbs.). These cattle graze on forages to gain weight before entering a feedlot.

The difference between stockers and feeders is their weight, age, and diet.

December comes calling in Virginia with frosty delight.
The trees stand tall and bare, their colorful leaves now a memory.
Mountains wear blankets of sparkling white.
The whole state hums with the magic of winter's first hello.

For the love of beef, a stocker producer in **Montgomery County**
has 100 stockers ready to go. They've left their moms,
grown up strong, munching on grass all season long.
Each stocker now weighs 700 to 900 pounds. **WOW** —
that's as much as a vending machine full of snacks!

So, off they go,
sold with pride, to a cattle buyer
in **Floyd County**
for the next big ride!

For the love of beef, look what's rolling in to Tazewell County — a neon green semi-truck! It's pulling a trailer, shiny and tall, with holes like polka dots. What's that called? A potbelly! Yes, that's the name, and hauling cattle is its game.

The driver shifts, then backs up slow... to the loading chute, where the stockers go. One by one, they step in line, each weighing 800 pounds.

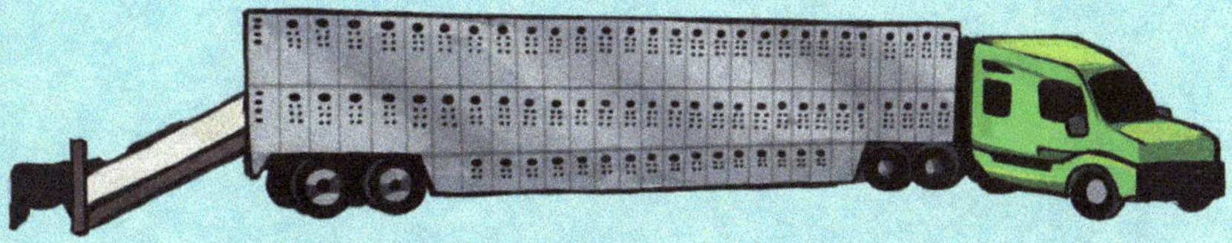

How much can that potbelly hold? 50,000 pounds. **WOW** — that's a load! The stockers climb inside until the trailer is full. Then, it's moo-ving on, ready for a Nebraska view.

Moo-vement Moment

Now shout it out: "Honk-Honk!"

BEEF UP YOUR KNOWLEDGE

Potbelly Trailer: a type of multi-deck livestock trailer that has a curved or dropped lower deck (belly)

- Designed to haul large numbers of livestock such as pigs or cattle
- Includes multiple decks, usually two or three with internal ramps
- Interior has numerous compartments divided by gates; a lowered deck reduces rollover risk
- Slats or punched-out holes provide air and light during transport
- Federal transportation laws limit the weight to 80,000 lbs., which generally refers to gross weight and includes the weight of the tractor, trailer, and the animal

For the love of beef, Virginia cattle have arrived at the Nebraska feedlot. And boy are they hungry! So, what's on the menu? Grains like corn and soybean meal — **Crunch, Crunch, Crack**!

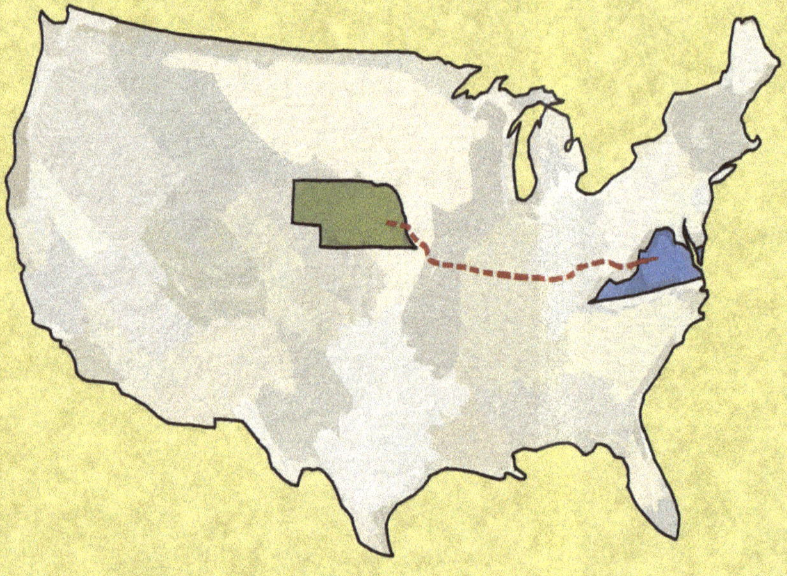

Feedlot cowboys, nutritionists, and veterinarians work together to keep the cattle healthy and happy — what a crew! The beef cattle munch and munch. They grow. They thrive. That's called "finishing."

BEEF UP YOUR KNOWLEDGE

Feeder Cattle: older and heavier cattle weighing 600–900 lbs. that are ready for the feedlot or already in a feedlot eating a high-energy grain diet to get fat

Feedlot = Feedyard: a place where beef cattle eat and grow before being processed for food

Most feedlots are located in the Midwest area of the United States. Iowa, Nebraska, and Kansas rank at the top for the most feedlots. Illinois and Minnesota are also part of the Midwest feedlot states. In addition, feedlots are found in the Eastern US, including Ohio and Pennsylvania.

Feedlots are in states that have an abundant corn supply and are near packing facilities.

For four to six months, they eat and play until they're big and strong. Then, it's time to check their size. The cattle step up, one by one — **Ker-Thunk!** goes the scale. **Beep-Beep!** flashes the number without fail. Whoa! Look at that size —

1,200 to 1,400 pounds! Not a surprise! Now that they're grown, there is no turning back. We've reached the final stop on our beef cattle adventure. Where do they go? To a place that is calm and neat called a packing plant, where they're prepared to become good food to eat. That's the beef life — from start to end.

BEEF UP YOUR KNOWLEDGE

Finishing: final stage of beef production designed to optimize body weight to achieve fat cover (finish) and intramuscular fat (marbling) for desired tenderness and flavor

Livestock Nutritionist: type of animal scientist who designs and manages feeding programs for farm animals

Packing Plant: Also called a processing facility, cattle are transported here once they reach market weight (1,200–1,400 lbs. at 18 to 22 months of age).

Inspectors from the United States Department of Agriculture (USDA) are on site to oversee processing, ensuring animal welfare and safety and quality standards are followed.

When processing is complete, beef products are sold and shipped in the United States and abroad to retail stores (grocery stores) and food service (restaurants).

Beef gives you strength and energy to run and play. Lean beef is a tasty treat, packed with protein — your muscles love it! It's full of zinc and iron, too, and B-vitamins that help you feel great.

From juicy burgers to sizzling steaks. To roasts that smell so good your tummy does a happy dance. Beef is on the plate!

Here's something cool:
A steer that weighs 1,000 pounds gives about 400 pounds of beef for families to enjoy. Almost the whole animal is used — about 99%.

So…what about the other 600 pounds?
Those parts make lots of things we use every day.
We call them by-products. Nothing goes to waste!

Leather boots. Baseball gloves and football gear that athletes love.
Hundreds of medicines — even insulin! Gelatin for ice cream
and marshmallows — a yummy treat! Even crayons, gum,
and soap, and toothpaste too — now that's no joke!

Moo-vement Moment

Now shout it out: "Beef is amazing!"

From calf to cow, from pasture to plate —
what an adventure we've had!
We've traveled across the beautiful
Commonwealth of Virginia,
visiting the top beef producing counties,
meeting farmers who care for their cattle through
winter, spring, summer, and fall, working hard all for
the love of Virginia beef.

Tip your cowboy hat one last time and give a big cheer
for Virginia farmers and their moo-velous friends!
Yee-Haw for Virginia beef!

THE END!

Moo-vement Moment

Say it loud: "Yee-Haw for Virginia beef!"

Data pulled in the calendar year 2025 from the United States Department of Agriculture's National Agricultural Statistics Service's (USDA-NASS) website (www.nass.usda.gov) provided the list of the top 25 beef producing counties in Virginia featured in *For the Love of Virginia Beef*. The counties featured were not listed in any specific order.

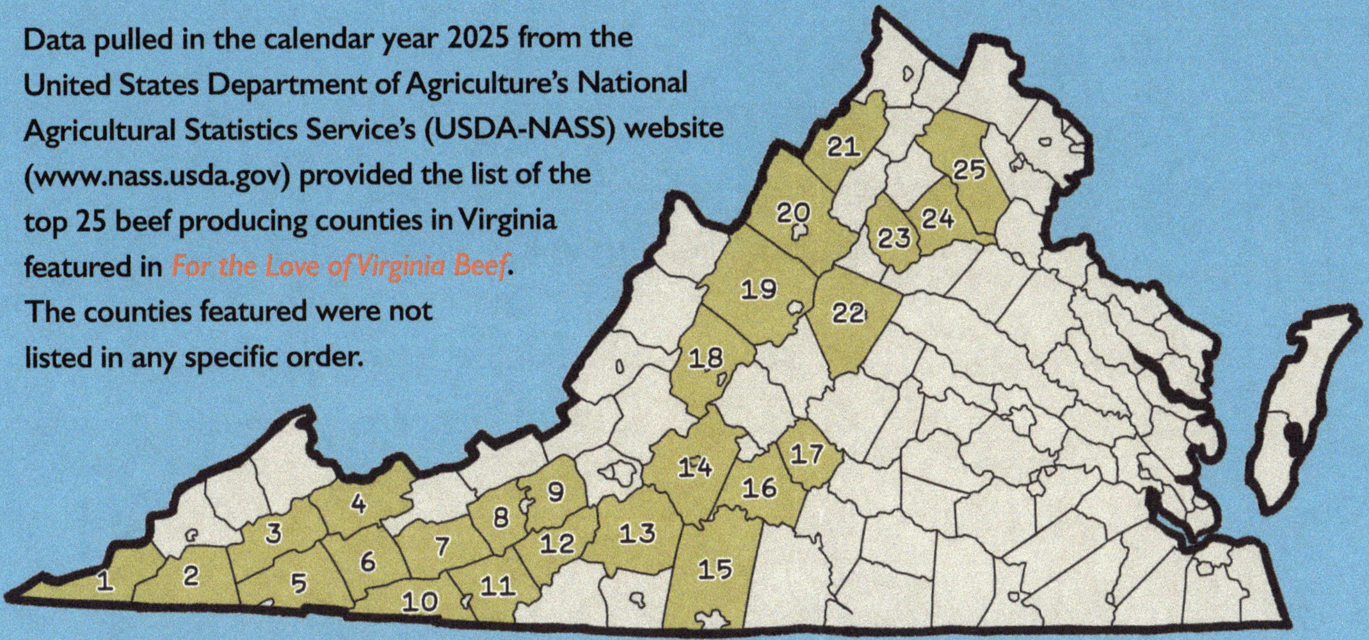

Want to know how many beef cattle are in your state or county?
Visit: www.nass.usda.gov/Data_and_Statistics/County_Data_Files?Livstock_County_Estimates/index.php

25 Featured Virginia Beef Producing Counties

1] Lee
2] Scott
3] Russell
4] Tazewell
5] Washington
6] Smyth
7] Wythe
8] Pulaski
9] Montgomery
10] Grayson
11] Carroll
12] Floyd
13] Franklin
14] Bedford
15] Pittsylvania
16] Campbell
17] Appomattox
18] Rockbridge
19] Augusta
20] Rockingham
21] Shenandoah
22] Albemarle
23] Madison
24] Culpeper
25] Fauquier

A Big MOO of Thanks

to the following for helping us share the story of Virginia's beef cattle industry.

From pasture to page, these individuals kept this project moo-ving forward.

Collin R. Ball, *Russell County, VA*

Bruce & Sylvia Warner, *Russell County, VA*

Pat Rasnake, *Russell County, VA*

Jennifer Fletcher Meade, *Scott County, VA*

Kip & Donna Kirby, *Tazewell County, VA*

Heather Musick, *Russell County, VA*

Janie Jessee, Jan-Carol Publishing, Inc.

FARM BUREAU RUSSELL

Russell County Farm Bureau Board of Directors and Staff

Russell County Farm Bureau Women's Leadership Committee

Letters of Support

Copenhaver Brothers Farms – Gene Copenhaver, President

First Bank & Trust Company – W. Mark Nelson, President/CEO

Leonard Land and Livestock – David A. Leonard, President

Meade Tractor – Kasey Kohl, Division President

Russell County Cattlemen's Association – Zac Ketron, Vice-President

Russell County, VA Public Schools – Dr. Kim Hooker, Division Superintendent

Russell County, VA Public Schools – Greg Mullins, Career and Technical Education Supervisor

Smyth/Washington Cattleman's Association – Lindsey Pratt, President

VA Cooperative Extension Russell County Office – Scott Jessee, ANR Extension Agent

Virginia Cattlemen's Association – Brandon Reeves, Executive Director

Financial Sponsors

Prime Sponsor: $1000+

VIRGINIA CATTLE INDUSTRY BOARD

This project is funded partially by the Virginia Cattle Industry Board.

Choice Sponsor: $500–$999

Leonard Land and Livestock – *Russell County, VA*

Virginia Cattlemen's Association – *Russell County, VA*

Select Sponsor: Up to $499

Ball Farms, LLC – *Russell County, VA*

Copenhaver Brothers Farms – *Washington County, VA*

First Bank & Trust Company

Louis White Farms & Livestock, LLC – *Russell County, VA*

Beefy Resources

For more information about agriculture, beef cattle, and educational resources, visit these organizations and websites:

- National Agriculture in the Classroom: agclassroom.org
- Virginia Agriculture in the Classroom: virginia.agclassroom.org
- National Cattlemen's Beef Association: ncba.org
- American National CattleWomen: ancw.org
- Virginia Beef Industry Council: vabeef.org
- Virginia Cattlemen's Association: vacattlemen.org
- Virginia Cooperative Extension: ext.vt.edu
- Virginia Department of Agriculture and Consumer Services: vdacs.virginia.gov/education-resources-for-teachers-and-students.shtml

About the Author

PHOTOGRAPHY BY KALYN, ROSEDALE, VA

Angela White Ball, PhD grew up on a beef cattle farm in Russell County, Virginia, considered the "heart of Southwest Virginia". This farm, a designated Virginia Century Farm, is still operated by her family today. Agriculture was deeply ingrained in Angela's life, shaping her path toward studying agricultural and applied economics and education at Virginia Tech. After college, she returned to Russell County to build a long career in public service supporting agriculture and natural resources.

Angela currently resides on Ball Farms, LLC in Russell County, where her family raises commercial beef cattle, hair sheep, and American Quarter Horses. She is a proud mom to her son Collin and an enthusiastic supporter of his many activities. Angela is also a strong advocate for agricultural literacy. When she isn't on the farm, she enjoys spending time on the beaches of South Carolina.

Angela wrote this book to showcase Virginia's cattle industry and the dedicated cattlemen and cattlewomen behind it, and to inspire children to learn how food makes its way from farm to table.